THE POETRY OF BERKELIUM

The Poetry of Berkelium

Walter the Educator

Silent King Books

SILENT KING BOOKS

SKB

Copyright © 2024 by Walter the Educator

All rights reserved. No part of this book may be reproduced in any manner whatsoever without written permission except in the case of brief quotations embodied in critical articles and reviews.

First Printing, 2024

Disclaimer
This book is a literary work; poems are not about specific persons, locations, situations, and/or circumstances unless mentioned in a historical context. This book is for entertainment and informational purposes only. The author and publisher offer this information without warranties expressed or implied. No matter the grounds, neither the author nor the publisher will be accountable for any losses, injuries, or other damages caused by the reader's use of this book. The use of this book acknowledges an understanding and acceptance of this disclaimer.

"Earning a degree in chemistry changed my life!"
- Walter the Educator

dedicated to all the chemistry lovers, like myself, across the world

BERKELIUM

Nestled within the periodic scroll,

BERKELIUM

Its secrets hidden, a mysterious goal,

BERKELIUM

Berkelium whispers in the void,

BERKELIUM

A saga of wonder to be enjoyed.

BERKELIUM

Born in the fiery heart of stars,

BERKELIUM

Forged in celestial furnaces afar,

BERKELIUM

It journeys through the cosmic expanse,

BERKELIUM

A fleeting glimpse, a cosmic dance.

BERKELIUM

In labs of science, it finds its place,

BERKELIUM

Under the gaze of the human race,

BERKELIUM

A synthetic creation, yet nature's kin,

BERKELIUM

With properties rare, a tale to spin.

BERKELIUM

Atomic number ninety-seven it claims,

BERKELIUM

In its nucleus, a symphony of flames,

BERKELIUM

Protons and neutrons in intricate array,

BERKELIUM

Binding together in a delicate sway.

BERKELIUM

Its electrons dance in orbits wide,

BERKELIUM

A dance of energy, a cosmic ride,

BERKELIUM

With valence shells and bonds to form,

BERKELIUM

In the alchemy of matter, it weaves its charm.

BERKELIUM

Berkelium, oh element divine,

BERKELIUM

In laboratories, your wonders shine,

BERKELIUM

A testament to human endeavor,

BERKELIUM

Unraveling mysteries, now and forever.

BERKELIUM

Yet in your depths, mysteries lie,

BERKELIUM

Invisible to the human eye,

BERKELIUM

For who can fathom the atomic dance,

BERKELIUM

The secrets held in quantum trance?

BERKELIUM

Perhaps in time, we'll come to know,

BERKELIUM

The depths of Berkelium's cosmic glow,

BERKELIUM

Unlocking secrets, revealing the truth,

BERKELIUM

In the never-ending quest for youth.

BERKELIUM

So let us marvel at Berkelium's grace,

BERKELIUM

In the vastness of time and space,

BERKELIUM

A symbol of knowledge, of human quest,

BERKELIUM

In the grand symphony, we are but guests.

BERKELIUM

In laboratories, minds collide,

BERKELIUM

Seeking the truths that time can't hide,

BERKELIUM

And in Berkelium, we find our muse,

BERKELIUM

A glimpse of the universe, in hues.

BERKELIUM

So here's to Berkelium, element rare,

BERKELIUM

In its mysteries, we find our share,

BERKELIUM

Of wonder, of awe, of cosmic delight,

BERKELIUM

In the dance of atoms, in the eternal night.

BERKELIUM

ABOUT THE CREATOR

Walter the Educator is one of the pseudonyms for Walter Anderson. Formally educated in Chemistry, Business, and Education, he is an educator, an author, a diverse entrepreneur, and he is the son of a disabled war veteran. "Walter the Educator" shares his time between educating and creating. He holds interests and owns several creative projects that entertain, enlighten, enhance, and educate, hoping to inspire and motivate you.

Follow, find new works, and stay up to date
with Walter the Educator™
at WaltertheEducator.com

www.ingramcontent.com/pod-product-compliance
Lightning Source LLC
LaVergne TN
LVHW020134080526
838201LV00119B/3864